LIBRARY OF
AWESOME ANIMALS

SQUIRREL MONKEY

By Colleen Sexton

BEARPORT
PUBLISHING

Minneapolis, Minnesota

Credits: Cover and title page, © webguzs/iStock; 3, © kjorgen/iStock; 4–5, © miroslav_1/Getty Images; 7, © Zocha_K/iStock; 8, © ViktorCap/iStock; 9, © miroslav_1/iStock; 11, © Rick Carlson/Design Pics Inc/Alamy; 12, © Helen E. Grose/Shutterstock; 13, © Pstedrak /Dreamstime; 14, © Terry Allen/Alamy; 15, © Cavan Images/Alamy; 16, © Dewin ' Indew/iStock, © ulkan/iStock, © AOTTORIO/Shutterstock; 17, © Marek Stuchlý/Alamy; 18, © michaklootwijk/YAY Media AS/Alamy; 19, © Phichak/Alamy; 20–21, © owngarden/iStock; and 23, © miroslav_1/iStock.

Bearport Publishing Company Product Development Team
President: Jen Jenson; Director of Product Development: Spencer Brinker; Senior Editor: Allison Juda; Editor: Charly Haley; Associate Editor: Naomi Reich; Senior Designer: Colin O'Dea; Associate Designer: Elena Klinkner; Product Development Assistant: Anita Stasson

Library of Congress Cataloging-in-Publication Data

Names: Sexton, Colleen A., 1967- author.
Title: Squirrel monkey / Colleen Sexton.
Description: Minneapolis, Minnesota : Bearport Publishing Company, [2023] | Series: Library of awesome animals | Includes bibliographical references and index.
Identifiers: LCCN 2022006988 (print) | LCCN 2022006989 (ebook) | ISBN 9798885091138 (library binding) | ISBN 9798885091206 (paperback) | ISBN 9798885091275 (ebook)
Subjects: LCSH: Squirrel monkeys--Juvenile literature.
Classification: LCC QL737.P925 S49 2023 (print) | LCC QL737.P925 (ebook) | DDC 599.8/52--dc23/eng/20220314
LC record available at https://lccn.loc.gov/2022006988
LC ebook record available at https://lccn.loc.gov/2022006989

For more information, write to Bearport Publishing, 5357 Penn Avenue South, Minneapolis, MN 55419. Printed in the United States of America.

Contents

Awesome Squirrel Monkeys! 4

A Forest Home 6

Small and Soft 8

Through the Trees 10

Troop Talk 12

Staying Safe 14

A Forest Feast 16

Baby Monkeys 18

Growing Up 20

Information Station 22

Glossary 23

Index 24

Read More 24

Learn More Online 24

About the Author 24

AWESOME
Squirrel Monkeys!

With a flick of its long tail, a little monkey is on the move. *SWISH!* Leaping from branch to branch, squirrel monkeys are awesome!

SQUIRREL MONKEYS ARE NAMED AFTER SQUIRRELS. BOTH ANIMALS ARE KNOWN FOR SCAMPERING THROUGH TREES.

A Forest Home

There are five kinds of squirrel monkeys that live in the **tropical** forests of Central America and South America. They make their homes in the thick and shady trees near rivers. While they occasionally go down to the forest floor, squirrel monkeys spend most of their lives up in the trees.

A TROPICAL FOREST **HABITAT** HAS MANY DIFFERENT KINDS OF TREES. SQUIRREL MONKEYS CAN LIVE IN ALMOST ANY OF THEM!

Small and Soft

Look at that squirrel monkey hanging onto a tree branch! The little animal has a round head with big eyes and big ears. Its soft, short fur is mostly golden-yellow and gray. And no one can miss a squirrel monkey's tail—it's longer than the animal's small body.

WHITE MARKINGS ON THE FUR AROUND THEIR EYES MAKE SQUIRREL MONKEYS LOOK LIKE THEY ARE WEARING MASKS.

Through the Trees

Squirrel monkeys are built for moving in forests! The quick animals use their strong legs to climb and jump from tree to tree. Long fingers and toes help them grip branches, and their tails keep them balanced. As they leap around, squirrel monkeys leave their **scent** behind for other monkeys.

TO LEAVE A SCENT, SQUIRREL MONKEYS RUB THEIR OWN PEE ON THEIR HANDS AND FEET.

Troop Talk

Squirrel monkeys usually live in **troops** of 50 to 300 animals. The many monkeys in these big groups make sounds to communicate with one another. They chatter when they meet. *CHIT-CHIRP!* If they sense danger, squirrel monkeys let out loud, high-pitched warnings. *SCREECH!*

AT NIGHT, MONKEYS IN A TROOP HUDDLE TOGETHER. THEY SLEEP SIDE BY SIDE, CURLING THEIR TAILS AROUND THEIR BODIES.

Staying Safe

Troops of squirrel monkeys work together to watch out for **predators**. Eagles and other birds may attack from above. Snakes slither up trees. Leopards and other wildcats creep across the forest floor. When an enemy gets too close, many monkeys may even work together to fight off the dangerous animal.

A harpy eagle

PEOPLE ARE A **THREAT**, TOO. SOME PEOPLE HUNT SQUIRREL MONKEYS.

A Forest Feast

Squirrel monkeys also work together to find food. They travel through the trees in small groups, always on the lookout for a meal. The monkeys munch on fruit and gobble up insects. **YUM!** Sometimes, they'll also eat seeds, leaves, flowers, nuts, or eggs.

SQUIRREL MONKEYS GET WATER FROM THE FRUIT THEY EAT AND BY DRINKING RAINWATER.

Baby Monkeys

While squirrel monkeys often work together in their troops, the **males** sometimes fight. They battle to decide who will **mate** with **females**.

About five months after mating, a female gives birth to one baby. The tiny monkey climbs onto its mother's back, grabs her fur, and wraps its tail around her body.

SQUIRREL MONKEY BABIES ARE BORN DURING RAINY MONTHS. ALL THE RAIN MEANS THERE WILL BE LOTS OF FRUIT TO EAT!

Growing Up

A baby squirrel monkey is helpless without its mother. For several months, the baby's only food is milk from its mother's body. The baby even rides on its mother's back. After about a year, the young monkey can live on its own, but it may still stay close to its mother. A few years later, the squirrel monkey can mate and have its own baby.

SQUIRREL MONKEYS LIVE FOR 10 TO 15 YEARS IN THE WILD.

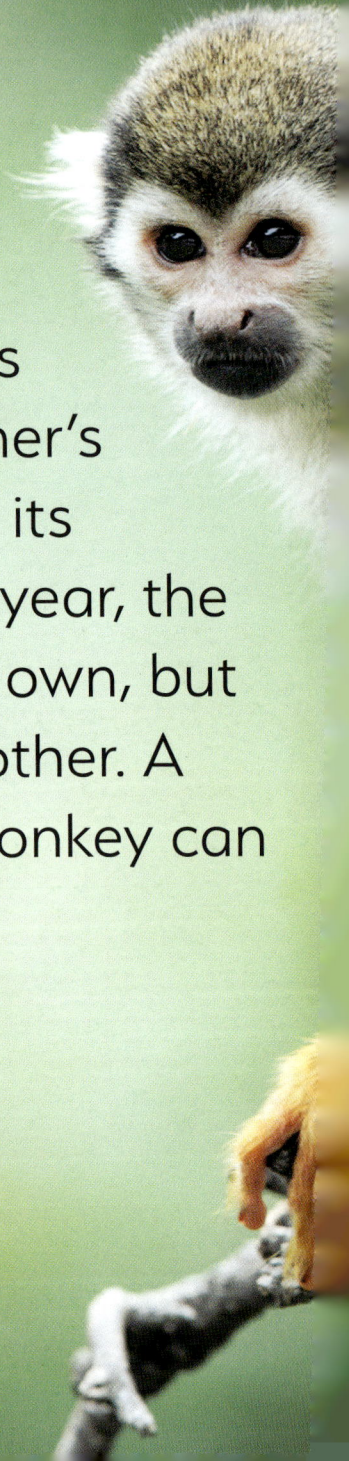

SQUIRREL MONKEYS ARE AWESOME!
LET'S LEARN EVEN MORE ABOUT THEM.

Kind of animal: Squirrel monkeys are mammals. Most mammals have fur, give birth to live young, and drink milk from their mothers as babies.

More monkeys: There are 264 types of monkeys. They live in South America, Africa, and Asia.

Size: The squirrel monkey's body is about 12 inches (30 cm) long. That's about the size of a small dog. But its tail is another 16 in. (41 cm) long.

SQUIRREL MONKEYS AROUND THE WORLD

Arctic Ocean

NORTH AMERICA

EUROPE

ASIA

Pacific Ocean

Atlantic Ocean

AFRICA

Pacific Ocean

N
W E
S

SOUTH AMERICA

Indian Ocean

AUSTRALIA

Southern Ocean

ANTARCTICA

WHERE SQUIRREL MONKEYS LIVE

females squirrel monkeys that can give birth to young

habitat a place where an animal lives

males squirrel monkeys that cannot give birth to young

mate to come together to have young

predators animals that hunt and kill other animals for food

scent a smell

threat something that might cause harm

troops large groups of squirrel monkeys that live together

tropical related to the hot, rainy places near Earth's equator

Index

babies 18–20, 22
ears 8
eyes 8, 16
feet 10
fruit 16–17, 19
fur 8, 10, 18, 22

hands 10
insects 16
predators 14
tails 4, 8, 10, 13, 18, 22
trees 5–6, 8, 10, 14, 16
troops 12–14, 18

Read More

Harris, Tim. *Wildlife Worlds South America (Wildlife Worlds).* New York: Crabtree Publishing, 2020.

Kenney, Karen Latchana. *Howler Monkeys (Blastoff! Readers: Animals of the Rain Forest).* Minneapolis: Bellwether Media, 2021.

Learn More Online

1. Go to **www.factsurfer.com** or scan the QR code below.
2. Enter "**Squirrel Monkey**" into the search box.
3. Click on the cover of this book to see a list of websites.

About the Author

Colleen Sexton is a writer and editor. She is the author of more than 100 nonfiction books for kids on topics ranging from astronauts to glaciers to sharks. She lives in Minnesota.